ANIMAL DEFENSES

HOW ANIMALS PROTECT THEMSELVES

WRITTEN BY ETTA KANER

ILLUSTRATED BY PAT STEPHENS

Kids Can Press

My thanks to Kimberly Baily and Bob Johnson of the Toronto Zoo for their helpful comments, to Pat Stephens for making the animals so lifelike, and to my editor, Laurie Wark, for her patience, persistence, flexibility and good humor.

For David

Text © 1999 Etta Kaner
Illustrations © 1999 Pat Stephens

All rights reserved. No part of this publication may be reproduced, stored in a retrieval system or transmitted, in any form or by any means, without the prior written permission of Kids Can Press Ltd. or, in case of photocopying or other reprographic copying, a license from CANCOPY (Canadian Copyright Licensing Agency), 1 Yonge Street, Suite 1900, Toronto, ON, M5E 1E5.

Kids Can Press acknowledges the financial support of the Ontario Arts Council, the Canada Council for the Arts and the Government of Canada, through the BPIDP, for our publishing activity.

Published in Canada by
Kids Can Press Ltd.
29 Birch Avenue
Toronto, ON M4V 1E2

Published in the U.S. by
Kids Can Press Ltd.
2250 Military Road
Tonawanda, NY 14150

Kids Can Press is a Nelvana company

Edited by Laurie Wark
Designed by Marie Bartholomew
Printed and bound in Hong Kong, China by Book Art Inc., Toronto

The hardcover edition of this book is smyth sewn casebound.
The paperback edition of this book is limp sewn with a drawn-on cover.

CM   99 0 9 8 7 6 5 4 3 2
CM PA   99 0 9 8 7 6 5 4 3 2

Canadian Cataloguing in Publication Data

Kaner, Etta
    Animal defenses

Includes index.
ISBN 1-55074-419-4 (bound)
ISBN 1-55074-421-6 (pbk.)

1. Animal defenses – Juvenile literature. 2. Animal Weapons – Juvenile literature. I. Stephens, Pat. II. Title.

QL759.K36 1999   j591.47   C98-932032-4

Contents

Introduction

What do you do when you are afraid? Do you yell for help? Do you hide? Do you run away? Some animals do these things too when they are afraid. But many animals defend themselves in more unusual ways. Some animals change color to make it hard for a predator, or enemy, to see them in their environment. An octopus can do this in seconds. Other animals pretend to be something they're not. An inchworm holds itself stiff to look like a stick. Some animals even have partnerships with other animals. Buffalo depend on birds to warn them of danger. And some crabs use an animal called an anemone like a sword. These are just a few of the strange ways in which animals defend themselves. Read on to find out more about the amazing world of animal defense.

Blue-ringed octopus in camouflage colors

Blue-ringed octopus

Putting on a show

Imagine that you are a very small animal. A bird is about to attack and it's too late to run. What do you do? You try to look dangerous, even though you're not. You hope that your new look will frighten away your enemy. Many harmless animals bluff like this in different ways.

Toad

Citrus swallowtail caterpillar

When a toad is cornered by a snake, it may puff itself up and stretch out its hind legs. The toad becomes about three times bigger than usual. This makes it look too big for the snake to swallow. If the toad is lucky, the snake slithers away to pick on someone its own size.

The citrus swallowtail caterpillar scares away hungry birds by acting like a snake. It lifts up the front of its body and sticks out a bright red organ that looks like a snake's forked tongue. It flickers the fake tongue back and forth and gives off a terrible smell. This is enough to turn birds away.

The blue-tongued skink of Australia is a slow-moving lizard. When frightened, it opens its mouth wide, hisses and sticks out a huge, bright blue tongue. It's not trying to be rude. It just wants to scare its attacker. Most hungry birds and mammals leave it alone. Wouldn't you?

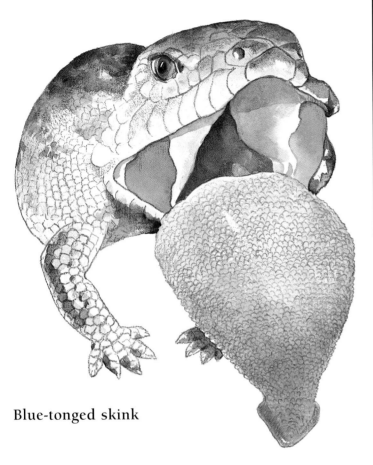

Blue-tonged skink

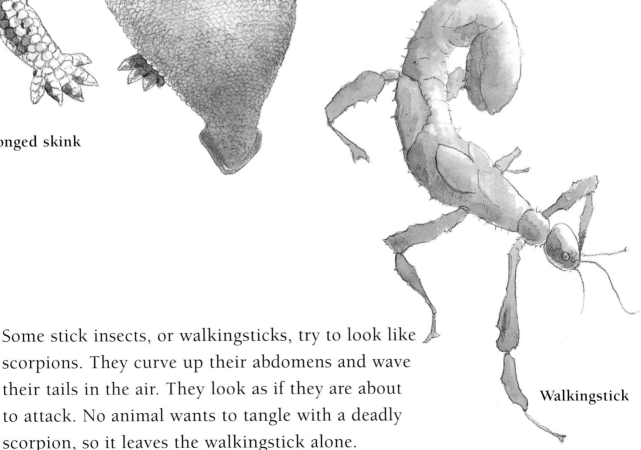

Walkingstick

Some stick insects, or walkingsticks, try to look like scorpions. They curve up their abdomens and wave their tails in the air. They look as if they are about to attack. No animal wants to tangle with a deadly scorpion, so it leaves the walkingstick alone.

Putting on a show

When resting on a branch, the eyed hawkmoth looks like a crumpled dead leaf. If a bird gets too close, the moth quickly uncovers its back wings. Suddenly, two huge eyespots stare out at the bird. They look as if they belong to an owl or a hawk. Now the bird wants to protect itself and it makes a fast exit.

Can you find me?

If an animal is not a fast mover, what does it do when a predator is close by? It hides. Some animals hide in their homes. Others don't go anywhere to hide. They just stay very still. Since they look a lot like their surroundings, their enemies don't see them. This is called camouflage.

Three-toed sloth

In the South American rain forest, this three-toed sloth spends most of its time hanging from a branch. It moves so little that tiny plants called algae grow on its long gray hairs. This makes the sloth look like the gray-green lichen plants that grow on branches around it. The sloth looks so much like a plant that some moths live in its fur.

10

The Australian tawny frogmouth sleeps in a tree all day long. Its brown feathers and the shape of its body make it look like a broken tree branch. At night, when the frogmouth hunts for insects, the darkness protects it.

Australian tawny frogmouth

Decorator crab

The decorator crab works hard to camouflage itself. With its claws it cuts pieces of seaweed from the clumps that lie on the seashore. It chews one end of the seaweed to make it sticky. Then it sticks the seaweed onto the bristles on its back. If the crab moves to a place with a different kind of seaweed, it changes its disguise.

11

Can you find me?

As the flounder hunts for food on the ocean floor, it rests on different backgrounds. If it sees a sandy seabed, the flounder's skin changes to look like sand. On a rocky bottom, its skin looks like tiny rocks. A flounder can even look like a checkerboard if it rests on one for a while.

Flounder

Prairie dogs live in burrows, or tunnels, under the ground in North America. A prairie dog can get into or out of its burrow through two or three holes. By standing on a mound of earth at one of these holes, a prairie dog keeps a lookout for enemies. As soon as it sees a hawk or a fox, it gives a warning bark. All of the prairie dogs in the area dive into their burrows and stay there until the coast is clear.

Prairie dog

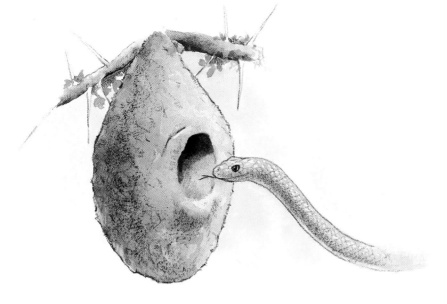

Nest of a Cape penduline tit

In South Africa, the nest of the Cape penduline tit has two entrances — a fake one and a real one. The fake one is a large hole in the side of the nest. If a snake enters it to steal the bird's eggs, surprise — it hits a wall. The real entrance is a tiny slit just above the false one. After the tit squeezes in or out, the real entrance closes up tight.

Chuckwalla

The chuckwalla doesn't have a home in which to hide. This lizard lives in a rocky American desert. When it sees a predator, the chuckwalla runs to the closest rock crevice and wedges itself in. Then it blows itself up with air. The chuckwalla's body becomes so tightly wedged between the rocks that it is impossible for the predator to pull it out.

Some snakes and lizards only half hide from their enemies. They wriggle their bodies into sand so that only the tops of their heads and their eyes are above the sand. By lying still, they can look out for danger and strike out at any passing meal.

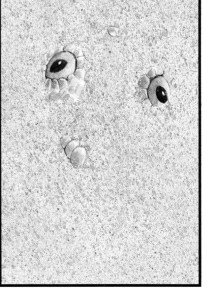

Can you find me?

A beaver's home, or lodge, is built from tree branches and mud. The dome-shaped lodge is built in a pond. Even though the lodge is in water, there is a dry, cozy room inside. The only way to get into this room is by diving deep and swimming up a tunnel that leads to it. A beaver has no trouble doing this. But there's no way a bear or a mountain lion can follow it.

Copycats

Is being a copycat always a bad thing? Not if you're an animal. In fact, that's how many animals survive. They look and act like animals that make their predators sick. So predators stay away from both them and their poisonous look-alikes.

Viceroy butterfly

Monarch butterfly

The viceroy butterfly looks like the monarch butterfly, doesn't it? Birds find it hard to tell them apart too. While the viceroy is harmless, the monarch is poisonous. Birds have learned to avoid eating both butterflies. They don't want to take a chance on getting ill by eating the wrong one.

Ant

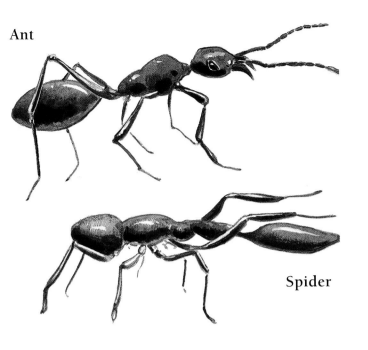

Spider

Many spiders copy ants. But it's not easy. Ants have six legs. Spiders have eight. Ants have antennae. Spiders don't. A spider must hold two legs in front of its head to look like antennae. It moves them from side to side and scurries around like an ant. Why do spiders imitate ants if they have to work so hard at it? Birds and lizards eat spiders, but they don't touch ants. They know that ants bite, sting and sometimes spray acid.

Hoverfly

Honeybee

The hoverfly looks and acts like a honeybee. It drinks nectar from flowers and buzzes a warning when threatened. But the hoverfly is harmless. It has no sting. Luckily, birds don't seem to know this. They are fooled into thinking that the hoverfly is a harmful honeybee, and leave it alone.

Copycats

At first glance, these two snakes look like twins. But they're not. The king snake is harmless, while the coral snake has a venomous bite. Animals can't tell the difference between them, so they stay clear of both. How can you tell the difference? This old saying might help, "If red touches yellow, avoid the fellow."

Coral snake

King snake

You can't hurt me

Just as you wear a helmet to protect your head, some
animals have protective gear too. But animals' gear is made
from shells, bone or spines, and can cover most of their body.

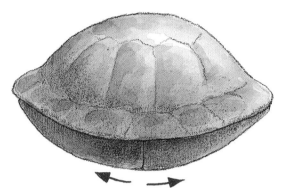

Box turtle

The three-toed box turtle has a top shell, or carapace, like all
turtles. But its bottom shell, or plastron, is different. It has a
hinge in the middle. This lets the turtle fold up the front and
back of its plastron. The top and bottom shells are sealed
together and the turtle's body is safe from any predator.

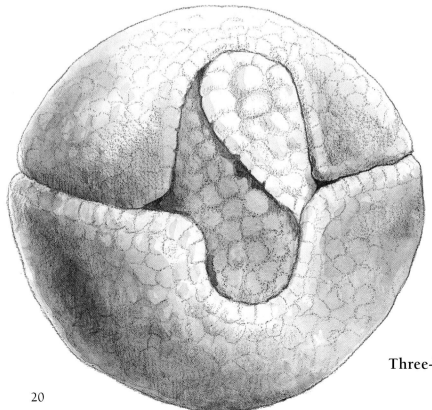

Three-banded armadillo

When the three-banded armadillo
from South America is attacked, it
rolls itself into a ball. Three bands
of hard shell on this mammal's
back protect its soft body inside.
If the predator touches it just after
it has rolled up, the armadillo
suddenly snaps the bands together
like a trap. Ouch! The animal's
nose or paw gets a nasty nip that
it will remember for a long time.

The hero shrew from West Africa well deserves this name. On the outside it looks like other shrews. On the inside, this tiny mammal has an unusually strong backbone. Its backbone is so strong that its small body can support a man standing on its back. If a person can't crush a hero shrew, could an animal possibly hurt it?

Hero shrew

Strange but true

Sometimes an armadillo will run away from its enemy. If it needs to cross a river, it either walks along the bottom or blows itself up like a balloon and paddles across.

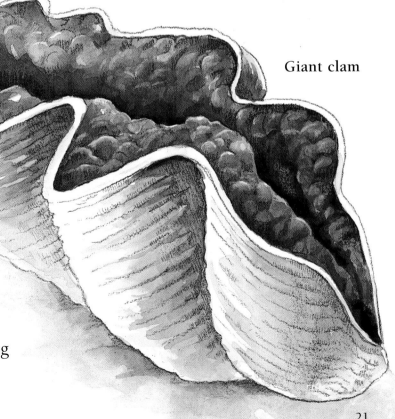

Giant clam

Some clam shells can be pried open by a starfish or smashed by a sea otter or drilled by a sea snail. Not the giant clam. Its shell is too thick and heavy. In fact, the giant clam has the largest shell in the world. It weighs about as much as three adults and is about as long as a bathtub.

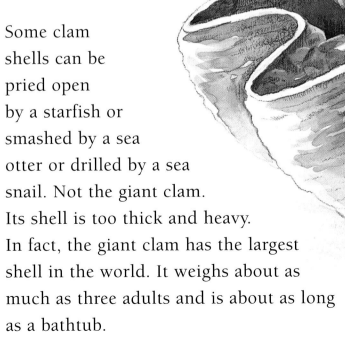

21

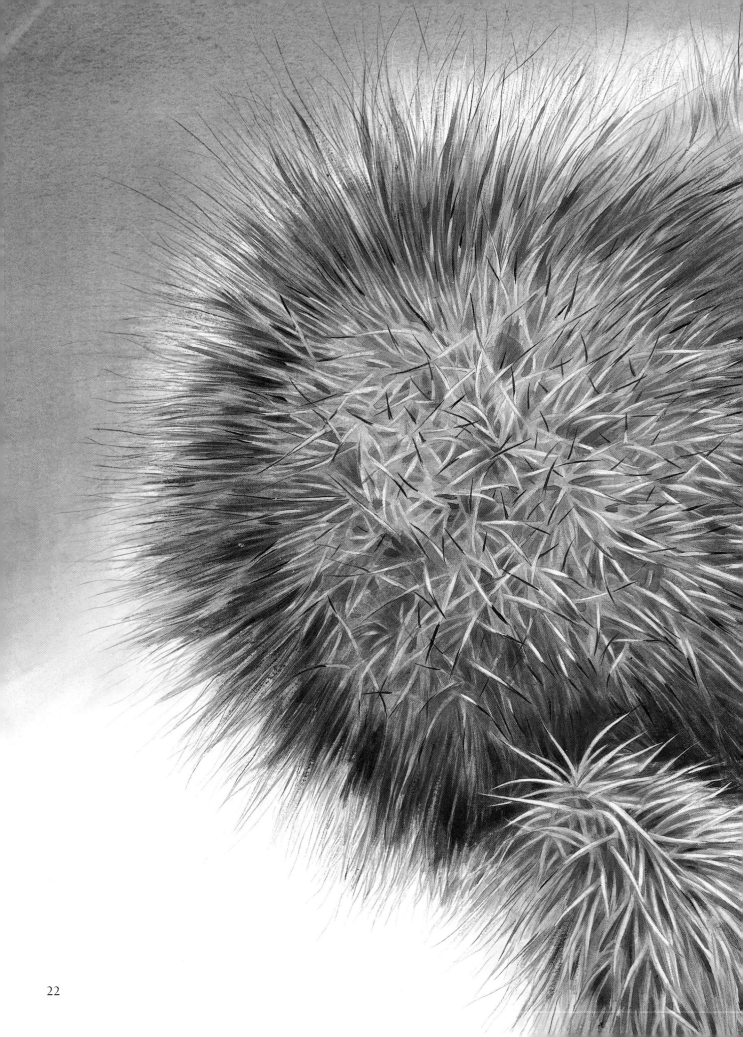

You can't hurt me

The North American porcupine's body is covered with about 30 000 quills. When the porcupine is threatened, it turns its back on its enemy. Then it lashes its tail wildly back and forth. If a fox or weasel is too close, it gets a snout full of quills. Can the fox pull them out? Not likely. The tip of each quill is covered with hooks, or barbs.

Warning, stay away!

Just as a red light tells you to stop, bright colors on animals warn predators to stop — and not eat them. If they do, they'll get very sick. Warning colors can be red or yellow or orange. They often form a pattern with black.

Even though the striped skunk hunts at night, its black and white marks are very clear. They mean "keep away." If a predator doesn't keep away, the skunk gives other warnings. It stamps its feet, arches its back and raises its tail. If the predator still doesn't get the hint, the skunk sprays it in the face. The spray is so strong that it blinds the predator for several hours.

Skunk

Cinnabar moth
caterpillar

A snake, bird or lizard wouldn't dare eat a cinnabar moth caterpillar. These animals know that if they did they would get very sick. The yellow and black stripes remind them of this. What makes this European caterpillar so poisonous? The ragwort plant that it eats has a poison in its leaves.

Ladybug

Depending on where it lives, a ladybug, or ladybird beetle, could have two, seven or twenty-two spots. The spots could be black on a red back, black on a yellow back, or red on a black back. No matter how many spots a ladybug has, birds, spiders and beetles don't bother it. They know that if they do, the ladybug's leg joints will give off a horrible-tasting liquid.

Strange but true

The Oriental fire-bellied toad has warning colors on its underside. When threatened, it twists its legs and arches its back to show its bright colors. These remind predators that the toad's skin contains a burning poison.

Warning, stay away!

These colorful poison-arrow frogs are easy to find in a South American rain forest. But they don't have to worry about being eaten. A bird or snake that tastes one immediately spits it out. Its skin is poisonous. The poison from one kind of poison-arrow frog can kill 50 people. No wonder animals usually stay clear of these frogs.

Let's stick together

Do you help your friends when they are in trouble? That's what animals do too. Some animals try to stay safe by living in groups. Others find a partner and the two of them help each other out. This is called symbiosis. Either way, they are safer from a predator than if they were alone.

Dolphins attacking a shark

Dolphins spend their whole lives in groups. When a shark threatens a baby dolphin, or calf, the group takes action. One or two dolphins swim in front of the shark to get its attention. As soon as the shark turns toward them, the other dolphins attack the shark from all sides. They hit the shark with their beaks until its gills are crushed and it drowns.

When an enemy breaks into a termite nest, soldier termites call for help. They can do this in two ways. They give off a scent that warns other termites of danger. They can also bang their heads on the walls of the nest. This sends vibrations through the nest which termites feel with their legs. When they get these messages, hundreds of termites rush to attack the invader.

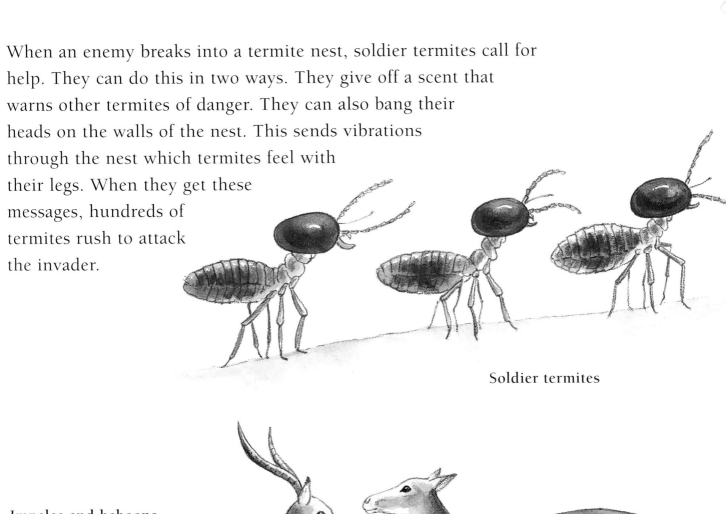

Soldier termites

Impalas and baboons

Groups of impalas and baboons often move around together on the plains. Impalas have excellent hearing and a keen sense of smell. Baboons have good eyesight. Together they keep watch for enemies. If they are attacked, the baboons are fierce fighters. This helps the impalas as well as the baboons.

Let's stick together

The hermit crab's partner is the sea anemone. The hermit crab lives inside a shell that has been cast off by another sea creature. The anemone lives on top of it. As the hermit crab looks for food along the seashore, the anemone gets a free ride. It also gets the leftovers from the crab's meals. In return, the anemone protects the crab with its fingerlike tentacles. The poisonous tentacles sting any animal that touches them.

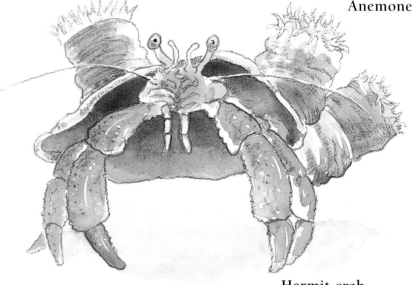

Anemone

Hermit crab

Oxpecker bird

African buffalo

The African buffalo has a personal alarm. It's the oxpecker bird. The oxpecker lives on the back of the buffalo. It eats the insects that burrow into the buffalo's hide. When the oxpecker senses danger, it screeches and flaps its wings. If the buffalo doesn't pay attention, the oxpecker uses its beak to rap the buffalo on the head. That usually gets the buffalo moving.

Shark

Remora

The remora is protected by sticking close to a shark. After all, who would attack a shark? The fish holds onto the shark with a large suction disc on its head. Wherever the shark swims, the remora goes with it, safe from enemies. In return, the remora cleans the shark's skin by eating the parasites that grow on it. It also eats the scraps of food the shark drops during its messy meals.

Strange but true

A little fish called Luther's goby and a blind shrimp are good partners. The shrimp digs a burrow for both to live in. When the goby guides the shrimp on feeding trips, the shrimp keeps its antennae in touch with the goby's tail. If there is danger, the goby wiggles its tail and the two of them hide in their burrow.

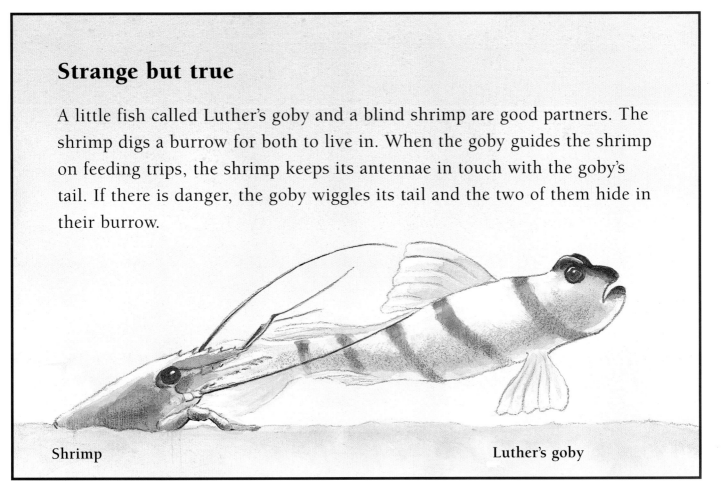

Shrimp

Luther's goby

Let's stick together

The clown fish is safe from enemies because it lives among the tentacles of a sea anemone. These tentacles are poisonous. Any fish that tries to catch a clown fish is stung to death by the anemone. It then becomes the anemone's dinner. Of course, the clown fish gets to share in the meal too. In exchange for protection, the clown fish chases away the anemone's enemies. It also keeps the anemone healthy by eating diseased parts.

Playing tricks

Many animals save their lives by playing tricks on their predators. No one teaches them to do these tricks. It just comes naturally.

Baby killdeer stay safe because their mother puts on an act when an enemy approaches. The killdeer mother flies away from the nest, then lands and drags one wing as if it is broken. The predator follows because a hurt bird is easy to catch. When the mother feels the nest is safe, she flies away leaving the predator behind.

Killdeer

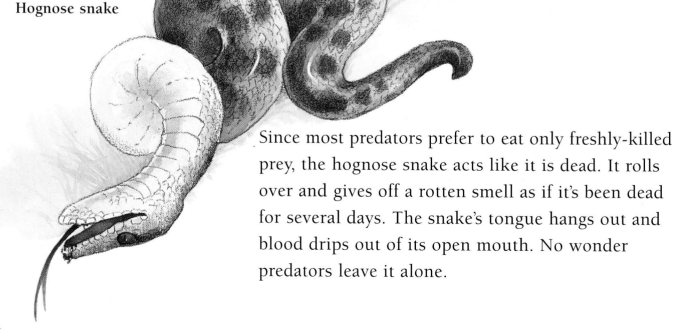

Hognose snake

Since most predators prefer to eat only freshly-killed prey, the hognose snake acts like it is dead. It rolls over and gives off a rotten smell as if it's been dead for several days. The snake's tongue hangs out and blood drips out of its open mouth. No wonder predators leave it alone.

Leopard gecko

Many lizards, like this Leopard gecko, have tails that break off when predators grab them. While the startled attacker watches the wriggling tail, the lizard has time to make a fast getaway. Amazingly, the lizard's tail grows back. But it's not as long or as straight as the original one.

Strange but true

Birds kill butterflies by attacking the head. But the hairstreak's head is hard to find. Hairstreaks have long fake antennae, which they wave about at the back of their wings. When a bird attacks these, the butterfly can usually escape.

Cuttlefish

A cuttlefish tries to blend in with its surroundings. When this doesn't work, it squirts a blue-black cloud into the water. A hungry shark attacks the inky blob, thinking it's the cuttlefish. Meanwhile, the cuttlefish quickly sneaks away.

Playing tricks

If the American opossum can't scare away its attacker, it "plays possum." The opossum tricks its enemy into thinking that it is dead. Suddenly, it falls over onto its side. Its tongue hangs out of its mouth and its eyes are half closed. Even if the predator pokes or bites it, the opossum doesn't move. It only moves again when it feels safe.

You can't catch me

Some animals are fast runners, fliers or swimmers. That's how they save their lives. But moving quickly is not always enough. Some animals have to outsmart their predators as well.

Pronghorn antelope

Pronghorn antelope can run fast over a long distance. But pronghorns don't just run away when they spot a wolf or coyote. They also give a warning signal by spreading out the long white hairs on their rumps. The large circles of hair reflect sunlight and can be seen from far away by other pronghorns. At the same time, the pronghorn gives off a strong scent which also sends a message that there are predators around.

Australian sugar-glider

It's hard for a predator to keep up with a sugar-glider. The Australian sugar-glider glides through the air from one tree to the next. It can travel as far as half a football field in a single glide. The sugar-glider has flaps of skin along the sides of its body connecting its hands and feet. When it glides, it stretches out the skin flaps and sails through the air. It steers with its fluffy tail.

When a red fox is being chased by a wolf or a lynx, it tries to hide its scent. It doesn't run in a straight line. It traces back over its own tracks. It runs through shallow water. It runs along the tops of fences and stone walls, and it even runs among cattle, pigs or deer. Once the predator loses the fox's scent, it can no longer chase the fox. The predator has been outfoxed!

Red fox

Index